Bibliographic information published by the German National Library:

The German National Library lists this publication in the National Bibliography; detailed bibliographic data are available on the Internet at http://dnb.dnb.de .

Imprint:

Print and binding: Books on Demand GmbH, Norderstedt Germany
ISBN: 9783668878808

This book at GRIN:

https://www.grin.com/document/454502

Oleg Gradov

Is it possible to consider ''cinetomics'' and ''argyromics'' as novel trends of functional morphology and systems biology of Protozoa?

GRIN Verlag

IS IT POSSIBLE TO CONSIDER «KINETOMICS» AND «ARGYROMICS» AS NOVEL TRENDS OF FUNCTIONAL MORPHOLOGY AND SYSTEMS BIOLOGY OF PARASITIC AND NONPARASITIC *PROTOZOA*? A BRIEF ANALYTICAL REVIEW

Oleg V. Gradov

Institute for Energy Problems of Chemical Physics, Russian Academy of Sciences; Institute of Biology and Chemistry, Moscow Pedagogical State University, Moscow, Russia

O.V. Gradov

Abstract. This paper considers argyrophilic (i.e. stained by the silver nitrate) fibrils visualized by the well-known Klein method and later called neuroformative system of *Protozoa*, or neuroformium. Klein described the excitation wave propagation and coordination of the cilia moving by this network, as well as the main role in localization of the peristomes and the ciliar apparatus on the topographic microanatomy map of *Protozoa* before and after the cell division. However, in early 1930-th the advanced staining techniques revealed two components in the above mentioned fibrill network with the similar morphobiochemical parameters (within the limits of the silver nitrate staining selectivity), one of which provides the mechanical stabilization of the *Protozoa* body (a number of *Paramecium* species were found to possess even a third selectively-stained fiber system). This resulted in the introduction of the term "argyrome" by Chatton in 1936 for the complex of the argentophilic fibrils and a term "cinetome" for the subpellicular infraciliature. It is therefore more appropriate to speak about "kinetomics" and "argyromics" rather than of a single "omics" for all the intracellular networks of *Protozoa*. We consider the principles of introduction of the novel "-omics" in modern science and from the above standpoint discuss whether it is possible to consider "kinetomics" and "argyromics" as the novel trends of functional morphology and systems biology of *Protozoa*.

Keywords: systems biology, kinetome, argyrome, *Protozoa*, *Ciliata*, "-omics" approaches, subpellicular infraciliature, neurophanes, neuromotor apparatus, neuroformative system, neuronemes, neuroformium, history of biology, history of protozoology.

Introduction: What is "omics"?

There is a general tendency in the contemporary cell biology to divide sub-disciplines in accordance with the morphological / ultrastructural and biochemical (selective staining) criteria. This classification considers the type of the compartments, organelles or subcellular localizations / morphobiochemical targets of the sighting study. This means that for any cytophysiological and ultrastructural unit one can introduce a corresponding sub-discipline, and hence, discover a "new field of science". A special name is given to every newly introduced branch of cell biology. This gives rise to the so-called "omic trend" in the cellular biology, because it is conventional to assign such areas and fields of research as "…-omics". The Oxford English Dictionary (also abbreviated as OED) distinguishes between the different meanings of the -ome suffix, including "a part of an animal or plant with a specified structure" for zoology and "all constituents considered collectively" for cellular and molecular biology. Consequently, it may be classified into the two types by the morphological / ultrastructural criteria (the first type) and biochemical selectivity (the second type). This classification will help us to understand the meaning of the applicability of new types of "omics" for protistology.

For example of morphological classification units see: membranomics for membranous structures (Ghosh, 2008; Shimanouchi, 2012); plastidomics or plastomics for plastids (Laukers, 2004; Navarro, 2004); chromonomics for chromosomes (Willard, 1999); mitochondriomics for redox-active interactions in chondriome (Reichert, 2004) and mitointeractomics for protein interactions in chondriome (Reja, 2009); cellomics (Taylor, 2007; Yasuda, 2012) and cytomics (Kriete, 2005; Valet, 2005; Schubert, 2006) for cells and cellular systems biology; histomics or tissomics for different living tissues (Ecker, 2005; Kriete, 2006); and, furthermore, phenomics – for phenotype and its physiological and morphological lability and reactivity (Furbank, 2011; Berger, 2012; Matsubara, 2012; Naika, 2013).

Many professional morphologists object to this classification. They point out that "no forward-looking biological meaning can be attached to this trend and to the new sciences", because "the results of our morphological investigation are not affected if our disciplines will be called as an "-omics" in a new fashion" and "strictly speaking, such a renaming is meaningless". But such a view is clearly a misunderstanding of the contemporary systems biology approaches and the methodology of the semiautomatic data mining in the computational biology and, consequently, in the peculiarities of the novel biological data which can not be processed using non-automatic, non-computational and non-mathematical research techniques. But novel complicated experimental techniques have made it possible and thus, we have a necessary and sufficient condition for an "omics trend" usability in biology (when other data processing methods are impossible or pointless). Nevertheless, there are many dangers in the use of the indiscriminate "omics" introduction into the general biology, when it is unreasonable in accordance with the above criteria. One source of confusion is the tendency for single structure emphasizing for any omic promotion without taking into account the relationships and interactions between different structures and processes in the living cell. This tendency contradicts to the system biological approach. Currently there is no satisfactory definition of systems biology, but we can now perform a more systematic study of the objects of different "omics" using multiple "omics" approaches (Zhang, 2010; Droste, 2011; Kayano, 2013; Kohl, 2014).

Consequently, to avoid any possibility of confusion we shall verify the morphological "omics" conclusions using biochemical "omics" antecedents. The best and the most complex conclusion for "omics" data mining is that obtained from the morphobiochemical data using colocalization studies and selective morphometry of spatial distributions and singular localizations of biochemical agents in cells. Early investigators used to ignore the spatial distributions of chemical agents in cells, but not in biochemical pathway network hierarchy. Hence, early systematizations of sub-disciplines according to "omic" hierarchy were performed in the frame of the organic chemical / biochemical principles. Such sub-disciplines (omics) were named on the basis of their chemical substrate implementations, such as: chemogenomics (Namchuk, 2002; Bredel, 2004; Wuster, 2008; Cheung-Ong, 2012; Wang, 2014; Brown, 2014), which is not fully equivalent to chemical genetics (Carlson, 2012; Jung, 2014); interferomics – a science about interfering RNAs (Reddy, 2013); lipidomics – a science about lipids and cellular fats (Lagarde, 2003; Kagan, 2004; Wenk, 2005); eicosanomics – a science about eicosanoids (Balazy, 2004); glycomics – a science about sugars (Kheronsky, 2003; Lubner, 2003; von der Lieth, 2004; Mrksich, 2004); ionomics – a science about ions in living cells (Salt, 2008; Baxter, 2009; Baxter, 2010; Danku, 2013); metalloproteomics – a science about metalloproteins (Shi, 2008; Barnett, 2012; Fu, 2014; Roberts, 2014) and similar metallomics as a science about the metal functions in living cells (Szpunar, 2004; Lopez-Barea, 2006; Szpunar, 2006; Mounicou, 2009); proteomics – a science about proteins (Geisow, 1998; Persidis, 1998; Müllner, 1998); kinomics – a science about enzymes-kinases (Johnson, 2005; Gomase, 2008; Vieth, 2005); phytochemomics – a science about phytochemical bioactive agents (Amigo-Benavent, 2014). Many principles of similar sciences can be automatically extended to the case of protistology – as complex principles for comparative colocalization morphology of the *Protozoa* species. In this paper we prove a generalization of this approach on *Protozoa* cytology / protistology using a conventional morphobiochemical approach based on the selective staining of the compartments of living cell.

Requirements on the protistological "omics"

Our first concern is to describe the systematic methods for distinction between the types of the stained microstructures in the frame of the "omic" approaches. Generally, there are various reactions for an accurate determination of different biochemical agents (as the objects of different "omic's" studies) in the living cells, based on the selective staining principles. For example, immunofluorescence can visualize and semi-quantitatively determine different types of biochemical agents using general coordination and supramolecular chemistry principles, such as non-covalent interactions and complementarity (the "lock-key" principle in its "antigen-antibody" implementation), etc. Similar determination methods are analogous in some aspects to the "mapping of the immunomic functions and the immunogenomic agents (or complements)" (Klysik, 2001; Brusic, 2003; Schönbach, 2003). This technique provides a rational basis on which the morphologists can make conclusions about different cell components. This approach can be extended by the simple general-biological interpolations to the *Protozoa* morphology not only because immunofluorescence staining methods are applicable for protistological / protozoological studies from 1960 till now (Zaman, 1965; Zaman, 1966; Dzbenski, 1966; Hauser, 1980; Kovac, 1980; Ohba, 1986; Brugerolle, 1988; Olins, 1989; Allen, 1990; Turkewitz, 1992; Dohra, 1994; Hanyu, 1995; Hanyu, 1996; Clerot, 2001; Santangelo, 2001; Strzyzewska-Jówko, 2003; McLaughlin, 2004; Itabashi, 2004; Kissmehl, 2004; Ramoino, 2004), but also because the simplest inorganic staining agents may be used for visualization of several compartments and constituents of the *Protozoa* cells (even without the antibodyome agents (Mohebbi, 2009; Zhu, 2013)). Notwithstanding, there is no case of determination of the different "omics" characteristics using inorganic dyes. In the microscopy practice, the investigator cannot always predict whether the staining agents are really specific for the stained compartments compositions or ROI (i.e. regions of interests) when using an inorganic dye without the "imbedded" molecular selectivity site(s). From the above we conclude that morphological "omics" based on the inorganic or bioinorganic determination (staining) agents can be used as the alternatives for the immunomics or other antibodyome-based dyes if and only if the visualization of the target compartments or biochemical constituents in the stained cell is equivalently effective both for inorganic-stained samples or living cells as well as for bioorganic-stained samples or living cells.

According to the hypothesis of colocalization microchemistry of the analyzed indicators, the cell components and the property / activity maps with the involved systems biology descriptors and pathways (Gradov, 2014; Orehov, 2014; Orehov, 2016) which form the basis of the "omics" targets, it is of great interest to consolidate the above listed descriptors using the principles of multimodal imaging morphochemistry (Vogler, 2010) and morphophysiology (Welsh, 1994). Thus one can speak about the morphochemistry-based "omic", when the physiological and biochemical basis of ROI is taken into account and considered in the frame of the methodologies and techniques of systems biology. A comprehensive approach must account for the integration of the biological hierarchy levels from proteins to compartments and organs / organoids in the frame of physiomics (Hunter, 2003; Gomase, 2008) for bio-mathematical model representations, implementations and interpretations based on the data mining principles (Britten, 2013). Also this approach must be modified in order to explain the fact that both internal and external regulation is the basis of physiological functions and, therefore, a quantitative imaging of ROI must be based on the quantitative mapping of the "metabo-regulome" (Okumoto, 2008) in the frame of regulomics of life as the form of existence of the protein bodies (Werner, 2003) (but without immunofluorescence this approach can not be applied for immunoregulomics (Tassinari et al., 2008)). However, these approaches do not explain the effects of the environmental agents and factors despite their significant and irremovable action on the morphology and physiology of *Protozoa*, as well as *Metazoa* (also as Plants). It also must be borne in mind that protistological objects are exposed to those factors and thus one can speak about the *Protozoa* exposome and the *Protozoa* exposomics (Buck Louis, 2012; van Tongeren, 2012; Miller, 2014; Vrijheid, 2014) at phylogenetic (including the local adaptations or / and microevolution) and genetic levels, including biobehavioral and geospatial (in the frame of population biogeography) aspects (Stahler, 2013; Lyon, 2014). But even at this stage we can project the phenomic phenomena of the "normal" physiomics modified by exogenous exposomic factors influence on the morphology of single cell or *Protozoa* organization using the analytic morphomics (Engelsbe, 2012; Englesbe, 2012a; Miller, 2014) and the timing-based analytic morphomic methodologies and approaches (Rinkinen, 2013; Ranganathan, 2014), interpolated from the human / animal (*Metazoa*) organization level to *Protozoa* considered both as the organisms and as single cells.

Which of the morphological and ultrastructural *Protozoa* compartments or organelles / organoids is most suitable for this novel "omics" as a proof of concept of the "morphophysiologically selective staining" (except from the objects of antibodyomics and immunomics)? The result of considerations depends on the fact that unicellular organisms in contrast to multi-cellular ones can not be analyzed by means of the standard "organomics", such as connectomics for neurosciences (Lichtman, 2008; Kennedy, 2010), epithelliomics (from "epitheliome") for covering tissues (Walker, 2004) etc., but their functional equivalents should inevitably be integrated into protistology, because the functionally corresponding structures really exist. Furthermore, it's generally known that some early classical protistologists developed a number of similar hypotheses / theories about the physiological correspondence between the compartments or organoids of the *Protozoa* cells and the multi-cellular organs of *Metazoa*. In accordance with the approach outlined in the first section of this article, we propose that each element of the *Protozoa* compartmentalization can be assigned to a one or more multi-cellular organs of *Metazoa* and each element can be characterized by two "omic" descriptors from morphology and biochemical physiology.

From kinetome and argyrome towards kinetomics and argyromics

The simplest example of such a hybrid morpho-physiological mapping and imaging data mining is the mapping of a pool of the contractile fibrills and conductive strands of *Protozoa*. The problem of existence of the special system of conductive fibrills in *Protozoa*, which are partially similar in their function to the nerve fibers of *Metazoa* and capable of coordination of the organoid activity in *Protozoa*, was comprehensively studied at the beginning of the XX century, although as early as in 1890 a detailed description of a myonem system in *Stentor* was given, which demonstrated a rapid reaction / response to different stimuli via a quick body contraction (Schuberg, 1890). The basis for the neural-like structure-coupled activity mapping was laid out at the beginning of the XX century when the novel staining method revealed a new system of fibrills – neurophanes which were expected to function as the conducting nerve fibers (Neresheimer, 1903). In *Stentor* they diverge at an angle and are parallel to the myonems, but differ from them in the affinity to different dyes, which allows to distinguish between them in the course of the mapping procedure. The same substances which are known to be the paralyzing agents for the nervous system of

Metazoa cause a similar effect in *Stentor,* since they affect the neurophanes. There is also a similarity between the neuromuscular relations in the topographic anatomy of higher animals: the neurophanes were shown to possess branches contacting with the myonems. Other *Protozoa,* such as *Diplodinium (Epidinium) ecaudatum* (a ciliate from the order *Enlodiniomorpha*) possess a fiber system, directed to the myonems as the moving organs. The above system was called motorium, and together with the fibers with the outcoming fiber bundle it was called a neuromotor apparatus (Sharp, 1914). The fiber bundle from the motorium goes to the point providing a sensitive character of the above apparatus. Sometimes there is also a number of additional centres (described at 1933 (Powers, 1933) for the first time) which sustain the functional activity of the general motorium.

Only motorium and micronucleus can be visualized by Mallory staining, although Mallory's trichrome stain (aniline blue + acid fuchsin + orange G) provides visualization of other cellular structures. A combination of the three different stains in precise amounts applied in the correct order reveals the structural details selectively. This results from more than just electrostatic interactions of the stain / dye with the tissue and the stain can not be washed out after each step. Collectively the stains complement one another (a so-called spectrozonal colocalization). We are already familiar with the fact that different Mallory staining method variations can be used for a large number of cellular constituents and tissues, different in their biochemical characteristics: keratin (Ayoub, 1963) and keratin aggregation products (Nan, 2006); inorganic agents, including heavy metals, such as lead and copper (Mallory, 1939); genetic molecules – RNAs (Chieffi Baccari, 1992); conductive nervous media components (Schümperli, 1977); electrostatic inert, inactive connective tissue matrixes (Cason, 1950) and secreting glands (Gude, 1953), etc. Thus, there is a close analogy between the classifications of various Mallory stained tissues and the possible "omics" for descriptor-based distinction and mapping-based differentiation of these tissues.

As an example of practical importance we consider a comparative projection of descriptors of the secreting cells (or tissues) to the subspace of the micro-colorimetric measurement results of the Mallory stained sample analysis in the frame of the secretome expression concept (Zwickl, 2005; Martinez, 2006) and comparative "secretomics" (Lamonica, 2005; Sibbald, 2006) which can also be used for the parasitic protozoa-related phenomena (Przyborski, 2004). A less trivial example is the relation between metallomics or microelementomics (Li, 2008) and an old Mallory method of lead and copper determination (Mallory, 1939), because to the best of our knowledge the methods of cellular metallomic and elementomic mapping have not yet been developed. A more intriguing example is the correlation between connectomics (Lichtman, 2008; Kennedy, 2010) and the topography of neurophanes (Neresheimer, 1903) and similar elements of the *Protozoa* body (as a simple example of the convergence or parallelism or pseudo-homology between *Protozoa* and *Metazoa* organizations), based on the semblances or propinquities of the Mallory staining results of the nervous media (Schümperli, 1977) and a neuromotor apparatus of *Protozoa*, as well as the relations between histomics or tissomics (Ecker, 2005; Kriete, 2006) of the synaptic neuromotor connections and the complexes of *Metazoa* and the neuromotor apparatus conjunctions or the branches of neurophanes, contacting with the myonems or spasmonems (Febvre, 1981). "The first point favoring the idea that the neuromotor apparatus is of a truly nervous character comes from animals stained with Mallory's connective-tissue stain" (Barrows, 1935). This illustrates our thesis about the necessity and inevitability of the development of new specialized "omic" for the systematic description of the *Protozoa* elements, functionally equivalent to the elements of the nervous "connectomics" of *Metazoa* or the neuromotor apparatus "physiomics".

If we introduce new terms for defining new research sub-branches, we cannot avoid multiple verification of our understanding of the subject under investigation, using alternative methods. Actually, it is not quite true to say that we have a new object of the analytic morphomics (Engelsbe, 2012; Englesbe, 2012a; Miller, 2014) if we use only one method for its structure determination. But the only method available to us from the foregoing text is Mallory staining. Although this is a reasonable visualization for our region of interest (ROI) as a morphomic descriptor, it would be wrong to postulate that it is not only necessary but also sufficient for the introduction of a novel entity, such as new "omics". It follows historically from the fact that the understanding of the neuromotor apparat, neurophanes and motoriums, since their first introduction have been sometimes criticized by a number of morphologists (Ten Kate, 1927; Ten Kate, 1928; Bretschneider, 1934) and then confirmed by the other ones (Fernandez-Galiano, 1949; Noirot-Timothée, 1958; Noirot-Timothée, 1960) in a different from the initial meaning and morphophysiological sense. Besides, we need purely inorganic methods of the structure visualization, since the selectivity of the most staining methods using the organic dyes can be attributed to the specific bioorganic interactions. Therefore the control staining method should avoid any biochemical interactions but it should also provide an effective visualization of the ROI.

An alternative approach to the visualization of the subpellicular microtubules or subpellicular infraciliature is either Ag or Au impregnation using the corresponding metal salts. This method was also developed at the beginning of the last century (Schuberg, 1905). Argyrophilic (i.e. stained by the silver nitrate) fibrils visualized by the well-known Klein method (Klein, 1926; Klein, 1927) and later called neuroformative system of *Protozoa*, or neuroformium (Klein, 1943), together form a kind of network similar to the connectome of *Metazoa*. Klein described the excitation wave propagation and coordination of the cilia moving by this network, as well as the main role in localization of the peristomes and the ciliar apparatus on the topographic microanatomy map of the *Protozoa* before and after the cell division. However, in early 1930-th the advanced staining techniques revealed two components in the above mentioned fibrill network with the similar morphobiochemical parameters (within the limits of the silver nitrate staining selectivity) (von Gelei, 1932; von Gelei, 1937), one of which provides the mechanical stabilization of the *Protozoa* body (a number of *Paramecium* species were found to possess even a third selectively-stained fiber system). This resulted in the introduction of the term "argyrome" by Chatton in 1936 for the complex of the argentophilic fibrils and a term "cinetome" for subpellicular infraciliature (Chatton, 1936; Chatton, 1937). It is therefore more appropriate to speak about "kinetomics" and "argyromics" rather that of a single "omics" for all intracellular networks of the *Protozoa*.

The initial idea of combination of all the argentophilic structures into the argyrome without considering the recent physiological, biochemical and ultrastructural data could lead to a simplified understanding of these structures as a mechanical network. According to the first electron microscopic observations of those structures (Metz, 1953; Sedar, 1955), the bundle fibrillar structure morphostructurally corresponds to the neuronemes or kinetodesmes, but there was no synapse-like connection observed between them in the pioneer works, while the other workers (Randall, 1958; Ehret, 1959; Yagiu, 1959) argued that many fibrils from several neighbouring kinetosomes can be connected to the kinetodesmic fibril. Coordinated motor activity of the cilia (Párducz, 1954) with the high level of reactivity to the electrostatic and electrodynamic sources and monovalent cations until the reverse movement of the cilia, provided by the fibrillar ectoplasmic structures with the system acetylcholine + cholinesterase localized within the ROI (Seaman, 1951; Seaman, 1951a; Seaman,1952), indicates the presence of a kind of similarity between some network structures in *Protozoa* and their neurophysiological / electrophysiological homologs and "hierarchical prototypes" of the high level connectome self-assembly in *Metazoa*. For a more detailed understanding of this discussion, see a fundamental Dogel's handbook (Dogel, 1965).

We do not present here a systematical development of this subject, but just propose a new approach and sub-disciplines ("omics"). A detailed discussion of such problems is beyond the scope of this paper, but we expect to publish our further results and considerations within several years. The structural changes of "argyrome" and "kinetome" ("cinetome" in French) induced by "regulomics" and "enviromics" are of special interest (Neiderhiser, 2001; Texiera, 2011). It is well known that "according to the A. Lwoff's (Lwoff, 1990) hypothesis the modifications of the argyrome pattern … are associated with the rearrangement of the cortical proteins" (Dovgal, 2002) and, priory, that "the ontogeny took place in any Ciliates and began from the moment of anlage the fragment of tomit kinetom" (Jankowski, 1972). Little is known about the role of the kinetome and argyrome morphogenesis in the formation and stabilization of the "neuromorphic" functions of *Protozoa*. The topology and topography of "kinetomes" undergo evolution in the course of the development and the adaptive changes in *Protozoa*. The topology of kinetomes was determined using the silver impregnation method (Chatton, 1930; Corliss, 1953) which used to be rather popular not only in France where it was initially developed, but also in Germany as "protargolimprägnation" based on the principle of the "protargolaffines" of the dried ciliates to the protargol, which results in the stained sample (protargolpräparaten) with the inactivated ciliates (Krainer, 1995). Despite the fact that the above method was widely used until recently in the CIS and Russian Federation (Alekperov, 2006), as well as in Asia and Africa (Quyen, 2014), the dry sample preparation by the Klein method described above is not a vital staining method. Only recently a number of woks have appeared describing the novel modifications of protargol impregnation staining for live observations (Hu, 2001). This previously existing limitation corresponds to the fact that Ag^+ ions and clusters in certain concentrations are very strong and active toxins and biocides. It should not be overlooked that "kinetome" is a term from "physiomics" (because dynamical elements, as well as kinetids are physiologically active), while "argyrome" is a biochemical and morphometrical (i.e., morphobiochemical) term from "morphomics". Consequently, "kinetomics" + "argyromics" = "dynamomics" + "localomics" according to the Parkinson systematics (Gradov, 2014). It is obvious that both dynamics and localization can be studied simultaneously only within the living object. Although vital Ag^+-based staining methods may affect the "dynamomics" leading to the artifacts, for the "localomics" studies non-vital methods produce much more artifacts due to the postmortem effects. Modern histochemical methods, such as FISH (fluorescence *in situ* hybridization) allow to study the kinetome and "kinetomics" *in vivo* (Modeo, 2013), but the full correspondence between the terms "kinetome" and "argyrom" in the old papers and in the recent works can be achieved only if the old staining methods using inorganic protargol staining on the one hand and the Mallory staining on the other hand) and the new visualization techniques (FISH and immunofluorescence) result in the colocalization in the ROI visualization (i.e. the ROI of the corresponding "omics").

Prospects of the development of the new "-omics"

With the development of the novel comparative system-biological (omical) approach the earlier classical papers considering kinetoms (cinetomes) and argyroms (argyromes) (for example, see: (Kaczanowski, 1970; Sapra, 1970; Ky, 1971; Wilbert, 1986; von Foissner, 1989; Wasik A, 1994) etc.) will still remain relevant and one will be able to revise them in the frame of the structure and function unity, i.e. both from the standpoint of statics and dynamics. At the same time the conventional silver staining method which has been standard for many years will also remain useful regardless of the further developments. This can be provided by the implementation of the alternative non-optical forms of the analytical signal registration. The simplicity and wide applicability of such methods as protargol impregnation staining and the Mallory method variations (Cason, 1950; Gude, 1953; Roque, 1953; Churg, 1956; James, 1961; Mazzugo, 1982; van Leeuwen, 1990) allows to expand the boundaries of the detection principles for the stained samples. A comprehensive analysis of the above possibilities based on the physical descriptor mapping (Orekhov, 2014; Orekhov, 2016), such the redox state (Gradov, 2013), determining the staining intensity through the medium acidity and the plasmon resonance on the small Ag clusters, is beyond the scope of this paper and will be partly considered in our forthcoming papers. Since mass-spectrometry is known to be a routine analytical method for the biochemical "omics" (Di Girolamo, 2013) with the highest accuracy in detection of the biochemical analytes, especially for the spin-labeled ones, this method is expected to be a final "diagnosis" for the introduction of kinetomics and argyromics. In this case both Ag and other inorganic non-immunochemical agents for morphological staining can be used as MS labels for colocalization analysis. The above fact is consistent with the existing analytical procedures where Ag^+ labeling is applied for characterization of hydrogen-bonded supramolecular assemblies by MALDI-TOF mass spectrometry (Timmerman, 2000), as well as the silver nanoparticles (Shrivas, 2008), while for ESI MS detection Ag-staining is useful as a complementary method for Coomassie-staining (Winkler, 2007), and for the MS analysis combined with gel-electrophoresis combined radiolabelling and silver staining can be successfully applied for visualization, localization, and identification of proteins or peptides (Westrook, 2001).

Due to their compatibility with the modern staining methods for MS and other "omics" methods, kinetomics and argyromics are expected to be useful not only in the morphological protistology, but also in the physiological and biochemical studies, such as classification and monitoring of the populations and communities of *Protozoa*, together with the other "omics" method – fingerprinting and the advanced analysis of the genetic polymorphism, introduced from genomics (Skotarczak, 2004; Przyboś, 2005; Zhang, 2006; Zhang, 2012).

References

1. Alekperov I, Buskey E, Snegovaya N. New and little known free-living ciliates from the plankton of the Caspian Sea. *Protistology* 2006, 4:195-201.
2. Allen RD, Ueno MS, Pollard LW, Fok AK. Monoclonal antibody study of the decorated spongiome of contractile vacuole complexes of *Paramecium*. *J Cell Sci* 1990, 96:469-475.
3. Amigo-Benavent M, Clemente A, Caira S, Stiuso P, Ferranti P, del Castillo MD. Use of phytochemomics to evaluate the bioavailability and bioactivity of antioxidant peptides of soybean β-conglycinin. *Electrophoresis* 2014, 35:1582-1589.
4. Ayoub P, Shklar G. A modification of the Mallory connective tissue stain as a stain for keratin. *Oral Surg Oral Med Oral Pathol Oral Radiol* 1963, 16:580-581.
5. Balazy M. Eicosanomics: targeted lipidomics of eicosanoids in biological systems. *Prostaglandins Other Lipid Mediat* 2004, 73:173-180.
6. Barnett JP, Scanlan DJ, Blindauer CA. Protein fractionation and detection for metalloproteomics: challenges and approaches. *Anal Bioanal Chem* 2012, 402:3311-3322.
7. Barrows AL. The Significance of skeletal variations in the genus Peridinium. *University of California Publications in Zoology* 1935, 18:397-478.
8. Baxter I. Ionomics: studying the social network of mineral nutrients. *Curr Opin Plant Biol* 2009, 12:381-386.
9. Baxter I. Ionomics: The functional genomics of elements. *Brief Funct Genomics* 2010, 9:149-156.
10. Berger B, de Regt B, Tester M. Trait dissection of salinity tolerance with plant phenomics. *Methods Mol Biol* 2012, 913:399-413.
11. Bredel M, Jacoby E. Chemogenomics: an emerging strategy for rapid target and drug discovery. *Nat Rev Genet* 2004, 5:262-275.
12. Bretschneider L. Beiträge zur Strukturlehre der Ophryoscoleciden. II. *Arch Protistenk* 1934, 82:298-330.
13. Britten RD, Christie GR, Little C, Miller AK, Bradley C, Wu A, Yu T, Hunter P, Nielsen P. FieldML, a proposed open standard for the Physiome project for mathematical model representation. *Med Biol Eng Comput*, 51:1191-1207.
14. Brown JB, Okuno Y, Marcou G, Varnek A, Horvath D. Computational chemogenomics: is it more than inductive transfer? *J Comput Aided Mol Des* 2014, 28:597-618.
15. Brugerolle G, Adoutte A. Probing protistan phylogenies with an anti-tubulin antibody. *Biosystems* 1988, 21:255-268.
16. Buck Louis GM, Sundaram R. Exposome: time for transformative research. *Stat Med* 2012, 31:2569-2575.
17. Carlson SM, White FM. Expanding applications of chemical genetics in signal transduction. *Cell Cycle* 2012, 11:1903-1909
18. Cason JE. A rapid one-step Mallory-Heidenhain stain for connective tissue. *Stain Technol* 1950, 25:225-226.
19. Chatton E, Lwoff A. Imprégnation, par diffusion argentique, de l'infraciliature des ciliés marins et d'eau douce, après fixation cytologique et sans dessication. *C R Soc Biol* (Paris) 1930, 104:834-836.

20. Chatton E, Lwoff A. Technique par l'étude des protozoaires, spécialement de leurs structures superficielles (cinetome et argyrome). *Bull Soc Franc Micr* 1936, 25-39.
21. Chatton E. Un nouvel élément de la structure des Sporozoaires: l'argyrome. *C R Acad Sci* (Paris) 1937, 204:633-637.
22. Cheung-Ong K, Song KT, Ma Z, Shabtai D, Lee AY, Gallo D, Heisler LE, Brown GW, Bierbach U, Giaever G, Nislow C. Comparative chemogenomics to examine the mechanism of action of dna-targeted platinum-acridine anticancer agents. *ACS Chem Biol* 2012, 7:1892-1901.
23. Chieffi Baccari G, Marmorino C, Minucci S, Di Matteo L, d'Istria M. Mallory stain may indicate differential rates of RNA synthesis: II. Comparative observations in vertebrate nuclei. *Eur J Histochem* 1992, 36:187-196.
24. Chieffi Baccari G, Marmorino C, Minucci S, Di Matteo L, Varriale B, d'Istria M, Chieffi G. Mallory stain may indicate differential rates of RNA synthesis: I. A seasonal cycle in the harderian gland of the green frog (Rana esculenta). *Eur J Histochem* 1992, 36:81-90.
25. Churg J, Prado A. A rapid Mallory trichrome stain (chromotrope-aniline blue). *AMA Arch Pathol* 1956, 62:505-506.
26. Clerot J, Iftode F, Budin K, Jeanmaire-Wolf R, Coffe G, Fleury-Aubusson A. Fine oral filaments in *Paramecium*: a biochemical and immunological analysis. *J Eukaryot Microbiol* 2001, 48:234-245.
27. Corliss JO. Silver impregnation of ciliated *Protozoa* by the Chatton-Lwoff technic. *Stain tech* 1953, 28:97-100.
28. Danku JM, Lahner B, Yakubova E, Salt DE. Large-scale plant ionomics. *Methods Mol Biol* 2013, 953:255-276.
29. Di Girolamo F, Lante I, Muraca M, Putignani L. The Role of Mass Spectrometry in the "Omics" Era. *Curr Org Chem* 2013, 17:2891-2905.
30. Dogel V.A. *General protozoology*. Oxford: Clarendon press; 1965.
31. Dohra H, Fujishima M, Fok AK, Allen RD. Monoclonal antibody to a bacterial endonuclear symbiont Holospora cross reacts with proteins of contractile vacuole radial canals of *Paramecium* species. *J Eukaryot Microbiol* 1994, 41:503-510.
32. Dovgal IV. The germ similarity in ontogeny of Ciliates (*Ciliophora*). *Vestnik Zoologii* 2002, 36(2):3-9.
33. Droste P, Miebach S, Niedenführ S, Wiechert W, Nöh K. Visualizing multi-omics data in metabolic networks with the software Omix: a case study. *Biosystems* 2011, 105:154-161.
34. Dzbenski TH. Immuno-fluorescent studies on Balantidium coli. *Trans R Soc Trop Med Hyg* 1966, 60:387-389.
35. Ehret C, Powers E. The cell surface of Pararneciuin. *Int Rev Cytol* 1959, 8:97-133.
36. Englesbe MJ. Quantifying the eyeball test: sarcopenia, analytic morphomics, and liver transplantation. *Liver Transpl* 2012, 18:1136-1137.
37. Englesbe MJ, Lee JS, He K, Fan L, Schaubel DE, Sheetz KH, Harbaugh CM, Holcombe SA, Campbell DA, Sonnenday CJ, Wang SC. Analytic morphomics, core muscle size, and surgical outcomes. *Ann Surg* 2012a, 256:255-261.
38. Febvre J. The Myomene of the Acantharia (*Protozoa*): a new model of cellular motility. Biosystems 1981, 14:327-336.
39. Fernandez-Galiano D. Sobre e aparato neurornotor y otras estructuras protoplasmicas de Opliryoscolex purkinjei Stein. *Trab Inst Cienc Nat Madrid*, Ser Biol 1949, 3:253-302.
40. Fu D, Finney L. Metalloproteomics: challenges and prospective for clinical research applications. *Expert Rev Proteomics* 2014, 11:13-19.
41. Furbank RT, Tester M. Phenomics – technologies to relieve the phenotyping bottleneck. *Trends Plant Sci* 2011, 16:35-644.
42. Geisow MJ. Proteomics: one small step for a digital computer, one giant leap for humankind. *Nat Biotechnol* 1998, 16:206.
43. Gelei G von. Ein neues Fibrillensystem im Ectoplasma von *Paramecium*; zugleich ein Vergleich zwischen dem neuen und dem alten Gittersystem. *Arch Protistenk* 1937, 89:133-162.

44. Gelei J von. Die reizleitenden Elemente der Ciliaten in nass hergestellten Silber - bzw. Goldpräparaten. *Arch Prolistenk* 1932, 77:152-174.
45. Ghosh D, Beavis RC, Wilkins JA. The identification and characterization of membranome components. *J Proteome Res* 2008, 7:1572-1583.
46. Gomase VS, Tagore S. Kinomics. *Curr Drug Metab* 2008, 9:255-258.
47. Gomase VS, Tagore S. Physiomics. *Curr Drug Metab* 2008a, 9:259-262.
48. Gradov OV. Experimental Setups for Ozonometric Microscopy. *Biomed Eng* 2013, 46:260-264.
49. Gradov OV, Gradova MA. Cryoelectron microscopy as a functional instrument for system biology, structural analysis and experimental manipulations with living cells. *Problems of Cryobiology and Cryomedicine* 2014, 24:193-210.
50. Gude WD. Modified Martins-Mallory stain for mouse pituitary gland. *Stain Technol* 1953, 28:161-162.
51. Hanyu K, Numata O, Takahashi M, Watanabe Y. Immunofluorescence localization of a 23-kDa Tetrahymena calcium-binding protein, TCBP-23, in the cell cortex. *J Biochem* 1996, 119:914-919.
52. Hanyu K, Takemasa T, Numata O, Takahashi M, Watanabe Y. Immunofluorescence localization of a 25-kDa *Tetrahymena* EF-hand Ca^{2+}-binding protein, TCBP-25, in the cell cortex and possible involvement in conjugation. *Exp Cell Res* 1995, 219:487-493.
53. Hauser M, Hausmann K, Jockusch BM. Demonstration of tubulin, actin and alpha-actinin by immunofluorescence in the microtubule-microfilament complex of the cytopharyngeal basket of the ciliate *Pseudomicrothorax dubius*. *Exp Cell Res* 1980, 125:265-274.
54. Hu X, Song W. Morphology and morphogenesis of *Holosticha heterofoissneri* n. sp. from the Yellow Sea, China (*Ciliophora, Hypotrichida*). *Hydrobiologia* 2001, 448:171-179.
55. Hunter PJ, Borg TK. Integration from proteins to organs: the Physiome Project. *Nat Rev Mol Cell Biol* 2003, 4:237-243.
56. Itabashi T, Terasaki T, Asai H. Novel nuclear and cytoplasmic proteins detected by anti-Zoothamnium arbuscula (*Protozoa*) spasmin 1 antibody in mammalian cells are dependent on the cell cycle. *J Biochem* 2004, 136:651-657.
57. James KR. A modified picro-Mallory method for use with formalin-fixed tissues. *J Med Lab Tech* 1961, 18:29-31.
58. Jankowski AW. Recapitulation of phylogenesis in ciliate ontogeny. *Problems of evolution* (Rus) 1972, 2:95-123.
59. Johnson SA, Hunter T. Kinomics: methods for deciphering the kinome. *Nat Methods* 2005, 2:17-25.
60. Jung DW, Kim WH, Williams DR. Chemical genetics and its application to moonlighting in glycolytic enzymes. *Biochem Soc Trans* 2014, 42:1756-1761.
61. Kaczanowski A. Morphological studies on Opalinids I. Staining and fragmentation of the pellicle of *Opalina ranarum*. *Acta Protozoologica* 1970, 7:205-210.
62. Kagan VE, Quinn PJ. Toward oxidative lipidomics of cell signaling. *Antioxid Redox Signal* 2004, 6:199-202.
63. Kayano M, Imoto S, Yamaguchi R, Miyano S. Multi-omics approach for estimating metabolic networks using low-order partial correlations. *J Comput Biol* 2013, 20:571-582.
64. Khersonsky SM, Ho CM, Garcia MA, Chang Y.T. Recent advances in glycomics and glycogenetics. *Curr Top Med Chem* 2003, 3:617-643.
65. Kissmehl R, Sehring IM, Wagner E, Plattner H. Immunolocalization of actin in *Paramecium* cells. *J Histochem Cytochem* 2004, 52:1543-1559.
66. Klein B. Ergebnisse mit einer Silbermethode bei Ciliaten. *Arch Protistenk* 1926, 56:243-279.
67. Klein B. Die Silberliniensysteme der Ciliaten. Ihr Verhalten während Teilung und Conjugation, neue Silberbilder, Nachtrage. *Arch Protistenk* 1927, 58:55-142.
68. Klein B. Das Silberlinien oder neuroformative System der Ciliaten. *Ann Nat Hist Mus Wien* 1943, 53:156-336.

69. Kohl M, Megger DA, Trippler M, Meckel H, Ahrens M, Bracht T, Weber F, Hoffmann AC, Baba HA, Sitek B, Schlaak JF, Meyer HE, Stephan C, Eisenacher M. A practical data processing workflow for multi-OMICS projects. *Biochim Biophys Acta* 2014, 1844:52-62.
70. Kovács P, Csaba G. Detection of histamine binding sites (receptors) in Tetrahymena by fluorescence technique. *Acta Biol Med Ger* 1980, 39:237-241.
71. Krainer KH. Taxonomische Untersuchungen an neuen und wenig bekannten planktischen Ciliaten (*Protozoa: Ciliophora*) aus Baggerseen in Österreich. *Lauterbornia* 1995, 21:39-68.
72. Kriete A. Cytomics in the realm of systems biology. *Cytometry* 2005, 68:19-20.
73. Ky H. New *Ciliata* from the intestine of freshwater fishes of Northern Vietnam. *Acta Protozoologica* 1971, 8:261-282.
74. Lagarde M, Géloën A, Record M, Vance D, Spener F. Lipidomics is emerging. *Biochim Biophys Acta* 2003, 1634:61.
75. Lamonica JM, Wagner M, Eschenbrenner M, Williams LE, Miller TL, Patra G, DelVecchio VG. Comparative secretome analyses of three Bacillus anthracis strains with variant plasmid contents. *Infect Immun* 2005, 73:3646-3658.
76. Laukens K, Witters E. Tobacco BY-2 Proteomics. *Biotech Agricult Forest* 2004, 53:332-343.
77. Li YF, Chen C, Qu Y, Gao Y, Li B, Zhao Y, Chai Z. Metallomics, elementomics, and analytical techniques. *Pure Appl Chem* 2008, 80:2577-2594.
78. López-Barea J, Gómez-Ariza JL. Environmental proteomics and metallomics. *Proteomics* 2006, 6:S51-S62.
79. Lubner GC. Glycomics: an innovative branch of science. *Boll Chim Farm* 2003, 142:50.
80. Lwoff A. Changements de structure et d'organisation des mailes du reseau cortical des cilies et allosterie. *C R Acad Sci Paris* 1990, 311:133-135.
81. Lyon DE, Starkweather AR, Montpetit A, Menzies V, Jallo N. A biobehavioral perspective on telomere length and the exposome. *Biol Res Nurs* 2014, 16:448-455.
82. Mallory FB. A contribution to staining methods: I. A differential stain for connective-tissue fibrillae and reticulum. II. Chloride of iron haematoxylin for nuclei and fibrin. III. Phosphotungstic acid haematoxylin for neuroglia fibres. *J Exp Med* 1900, 5:15-20.
83. Mallory FB, Parker F. Fixing and staining methods for lead and copper in tissues. *Amer J Pathol* 1939, 15:517-522.
84. Martinez G, Georgas K, Challen GA, Rumballe B, Davis MJ, Taylor D, Teasdale RD, Grimmond SM, Little MH. Definition and spatial annotation of the dynamic secretome during early kidney development. *Dev Dyn* 2006, 235:1709-1719.
85. Matsubara S, Förster B, Waterman M, Robinson SA, Pogson BJ, Gunning B, Osmond B. From ecophysiology to phenomics: some implications of photoprotection and shade-sun acclimation in situ for dynamics of thylakoids *in vitro*. *Philos Trans R Soc Lond B Biol Sci* 2012, 367:3503-3514.
86. Mazzucco G, Basolo B, Monga G. The use of Mallory's phosphotungstic acid-hematoxilin (PTAH) stain in renal pathology. *Pathol Res Pract* 1982, 175:380-391.
87. McLaughlin NB, Buhse HE. Localization by indirect immunofluorescence of tetrin, actin, and centrin to the oral apparatus and buccal cavity of the macrostomal form of *Tetrahymena vorax*. *J Eukaryot Microbiol* 2004, 51:253-257.
88. Metz Gh, Pitelka D, Westfall J. The fibrillar systems of Ciliates as revealed by the electron microscope. I. Paramecium. *Biol Bull* 1953, 104:408-425.
89. Miller AL, Min LC, Diehl KM, Cron DC, Chan CL, Sheetz KH, Terjimanian MN, Sullivan JA, Palazzolo WC, Wang SC, Hall KE, Englesbe MJ. Analytic morphomics corresponds to functional status in older patients. *J Surg Res* 2014, 192:19-26.
90. Miller GW, Jones DP. The nature of nurture: refining the definition of the exposome. *Toxicol Sci* 2014, 137:1-2.
91. Modeo L, Fokin SI, Boscaro V, Andreoli I, Ferrantini F, Rosati G, Verni F, Petroni G. Morphology, ultrastructure, and molecular phylogeny of the ciliate Sonderia vorax with insights into the systematics of order Plagiopylida. *BMC Microbiology* 2013, 13(40):1-23.

92. Mohebbi N, Wagner CA. The "antibodyome" or, how to find antibodies? *J Nephrol* 2009, 22:439-441.
93. Mounicou S, Szpunar J, Lobinski R. Metallomics: the concept and methodology. *Chem Soc Rev* 2009, 38:1119-1138.
94. Mrksich M. An early taste of functional glycomics. *Chem Biol* 2004, 11:739-740.
95. Müllner S, Neumann T, Lottspeich F. Proteomics – a new way for drug target discovery. *Arzneimittelforschung* 1998, 48:93-95.
96. Naika M, Shameer K, Sowdhamini R. Comparative analyses of stress-responsive genes in Arabidopsis thaliana: insight from genomic data mining, functional enrichment, pathway analysis and phenomics. *Mol Biosyst* 2013, 9:1888-1908.
97. Namchuk M. Finding the molecules to fuel chemogenomics. Targets 2002, 1: 125-129.
98. Nan L, Dedes J, French BA, Bardag-Gorce F, Li J, Wu Y, French SW. Mallory body (cytokeratin aggresomes) formation is prevented in vitro by p38 inhibitor. *Exp Mol Pathol* 2006, 80:228-240.
99. Navarro FA. Minidiccionario crítico de dudas. *Panace* 2004, 5:17-18.
100. Neiderhiser JM. Understanding the roles of genome and envirome: methods in genetic epidemiology. *Br J Psychiatry Suppl* 2001, 40:S12-S17.
101. Neresheimer E. Ueber die Höhe histologischer Differenzierung bei hete-rotrichen Ciliaten. *Arch Protistenk* 1903, 2:305-324.
102. Noirot-Timothée C. Etude au microscope électronique des fibres rétro-ciliaires des Ophryoscolecidae: leur ultrastructure, leur insertion, leur rôle possible. *C R Acad Sci* (Paris) 1958, 246:1286-1289.
103. Noirot-Timothée C. Etude d'une famille de Ciliés: les Ophryoscolecidae. Structure et ultrastructure. *Ann Sci Nat Zool* 1960, 1:331-337.
104. Ohba H, Ohmori I, Numata O, Watanabe Y. Purification and immunofluorescence localization of the mutant gene product of a Tetrahymena cdaA1 mutant affecting cell division. *J Biochem* 1986, 100:797-808.
105. Okumoto S, Takanaga H, Frommer WB. Quantitative imaging for discovery and assembly of the metabo-regulome. *New Phytol* 2008, 180:271-295.
106. Olins DE, Olins AL, Robert-Nicoud M, Jovin TM, Wehland J, Weber K. Differential distribution of alpha-tubulin isotypes in Euplotes eurystomus determined by confocal immunofluorescence microscopy. *Biol Cell* 1989, 66:235-246.
107. Orehov FC, Gradov OV. On-line Compatibility of COBAC, QSPR / QSAR and SBGN Technologies: the Unity of Theory and Practice for Biomedical Equipment Design and Biochemical Diagnostic Data Analysis. RGC on Biomed Eng, X, 2014. DOI: 10.13140/2.1.1616.1921.
108. Orehov FC, Gradov OV. Hybridization of COBAC, QSPR / QSAR and SBGN Technologies: the Unity of Theory and Practice for Biomedical Equipment Design and Biochemical Diagnostic Data Analysis. *J Med Bioeng* 2016, 5:128-132.
109. Persidis A. Proteomics. *Nat Biotechnol* 1998, 16:393-394.
110. Powers P. Studies on the Ciliates from sea urchins. II. Entodiscus borealis (Hentschel) (*Protozoa*, *Ciliata*), behaviour and morphology. *Biol Bull* 1933, 65:122-136.
111. Przyborski J, Lanzer M. Parasitology. The malarial secretome. *Science* 2004, 306:1897-1898.
112. Przyboś E, Tarcz S. Molecular polymorphism of strains within *Paramecium septaurelia* (*Ciliophora, Oligohymenophorea*). *Folia Biol* (Krakow) 2005, 53:87-93.
113. Ramoino P, Scaglione S, Diaspro A, Beltrame F, Fato M, Usai C. GABAA receptor subunits identified in *Paramecium* by immunofluorescence confocal microscopy. *FEMS Microbiol Lett* 2004, 238:449-453.
114. Randall J, Jackson S. Fine structure and function in Stentor polymorphus. *J Biophys Biochem Cytol* 1958, 4:807-830.

115. Ranganathan K, Terjimanian M, Lisiecki J, Rinkinen J, Mukkamala A, Brownley C, Buchman SR, Wang SC, Levi B. Temporalis muscle morphomics: the psoas of the craniofacial skeleton. *J Surg Res* 2014, 186:246-252.
116. Reddy AB. Genome-Wide Analyses of Circadian Systems. *Handbook of Experimental Pharmacology* 2013, 217:379-388.
117. Reichert AS, Neupert W. Mitochondriomics or what makes us breathe. *Trends Genet* 2004, 20:555-562.
118. Reja R, Venkatakrishnan AJ, Lee J, Kim BC, Ryu JW, Gong S, Bhak J, Park D. MitoInteractome: mitochondrial protein interactome database, and its application in 'aging network' analysis. *BMC Genomics* 2009, 10(Suppl3):S20.
119. Rinkinen J, Terjimanian M, Benedict M, Hiltzik D, Seyi A, Lisiecki J, Wang SC, Buchman SR, Levi B. Temporal morphomics as a model for determining preoperative risk of blood transfusion in nonsyndromic craniosynostosis patients. *Plast Reconstr Surg* 2013, 132:403e-412e.
120. Roberts EA, Sarkar B. Metalloproteomics: focus on metabolic issues relating to metals. *Curr Opin Clin Nutr Metab Care* 2014, 17:425-430.
121. Roque AL. Chromotrope aniline blue method of staining Mallory bodies of Laennec's cirrhosis. *Lab Invest* 1953, 2:15-21.
122. Salt DE, Baxter I, Lahner B. Ionomics and the study of the plant ionome. *Annu Rev Plant Biol* 2008, 59:709-733.
123. Santangelo G, Bruno P. An immunofluorescence technique for staining ciliated protozoans: highlighting cytoplasmic microtubular arrays and stages of micronuclear meiosis. *Micron* 2001, 32:207-210.
124. Sapra GS, Dass CMS. The cortical anatomy of *Stylonychia notophora* Stokes and morphogenetic changes during binary fission. *Acta Protozoologica* 1970, 7:193-204.
125. Schuberg A. Zur Kenntnis der Stentor coeruleus. Zool Jahrb Abt Anat 1890, 4:197-238.
126. Schuberg A. Über Cilien und Trichocysten einiger Infusorien. *Arch Protistenk* 1905, 6:61-110.
127. Schubert W. Cytomics in characterizing toponomes: towards the biological code of the cell. *Cytometry A* 2006, 69:209-211.
128. Schümperli RA. Adaptation of the Mallory-Azan staining method to insect nervous tissue. *Stain Technol* 1977, 52:55-56.
129. Seaman G, Houliham R. Enzyme systems in Tetrahymena geleii S. II. Acetilcholinesterase activity. Its relation to motility of the organism and to coordinated ciliars action in general. *J Cell Physiol* 1951, 37:309-321.
130. Seaman G. Localization of acetylcholinesterase activity in the protozoan Tetrahymena geleii S. *Proc Soc Exper Biol Med* 1951a, 76:169-170.
131. Seaman G. Enzyme systems in Tetrahymena geleii S. III. Aerobic utilization of hexoses. *J Biol Chem* 1951b, 191:439-446.
132. Seaman G. Studies on Protozoan metabolism and their relation to general problems of physiology and biocherhistry. *Tex Rep Biol Med* 1951c, 9:171-179.
133. Seaman G. The phosphagen of *Protozoa*. *Biochim Biophys Acta* 1952, 9:693-696.
134. Sedar A, Porter K. The fine structure of cortical components of *Paramecium multimicronucleatum*. *J Biophys Biochem Cytol* 1955, 1:583-604.
135. Sharp R. Diplodinium ecaudatum with an account of its neuromotor apparatus. *Univ Calif Publ Zool* 1914, 13:42-122.
136. Shi W, Chance MR. Metallomics and metalloproteomics. *Cell Mol Life Sci* 2008, 65:3040-3048.
137. Shimanouchi T. Membranomics Research Using Liposome Immobilization Technique. *Biophysics (Seibutsu Buturi)* 2012, 52(3):154-155.
138. Shrivas K, Wu HF. Applications of silver nanoparticles capped with different functional groups as the matrix and affinity probes in surface-assisted laser desorption/ionization time-of-

flight and atmospheric pressure matrix-assisted laser desorption/ionization ion trap mass spectrometry for rapid analysis of sulfur drugs and biothiols in human urine. *Rapid Commun Mass Spectrom* 2008, 22:2863-2872.
139. Sibbald MJ, Ziebandt AK, Engelmann S, Hecker M, de Jong A, Harmsen HJ, Raangs GC, Stokroos I, Arends JP, Dubois JY, van Dijl JM. Mapping the pathways to staphylococcal pathogenesis by comparative secretomics. *Microbiol Mol Biol Rev* 2006, 70:755-788.
140. Skotarczak B, Przyboś E, Wodecka B, Maciejewska A. 2004. Random amplified polymorphic DNA fingerprinting as a marker for *Paramecium jenningsi* strains. *Folia Biol* (Krakow), 52:117-124.
141. Stahler GJ, Mennis J, Baron DA. Geospatial technology and the "exposome": new perspectives on addiction. *Am J Public Health* 2013, 103:1354-1356.
142. Strzyzewska-Jówko I, Jerka-Dziadosz M, Frankel J. Effect of alteration in the global body plan on the deployment of morphogenesis-related protein epitopes labeled by the monoclonal antibody 12G9 in *Tetrahymena thermophila*. *Protist* 2003, 154:71-90.
143. Szpunar J. Metallomics: a new frontier in analytical chemistry. *Anal Bioanal Chem* 2004, 378:54-56.
144. Szpunar J. Advances in analytical methodology for bioinorganic speciation analysis: metallomics, metalloproteomics and heteroatom-tagged proteomics and metabolomics. *Analyst* 2005, 130:442-465.
145. Tassinari OW, Aponte M, Caiazzo RJ, Liu BCS. Immunoregulomics. *Methods Mol Biol* 2008, 441:163-173.
146. Taylor DL. Past, present, and future of high content screening and the field of cellomics. *Methods Mol Biol* 2007, 356:3-18.
147. Teixeira AP, Dias JM, Carinhas N, Sousa M, Clemente JJ, Cunha AE, von Stosch M, Alves PM, Carrondo MJ, Oliveira R. Cell functional enviromics: unravelling the function of environmental factors. *BMC Syst Biol* 2011, 92:1-16.
148. Ten Kate C. Über das Fibrillensystem der Ciliaten. *Arch Protistenk* 1927, 57:362-426.
149. Ten Kate C. Über das Fibrillensystem der Ciliaten. II. Das Fibrillensystem der Isotrichen (Isotricha und Dasytricha). *Arch Protistenk* 1928, 62:328-354.
150. Timmerman P, Jolliffe KA, Calama MC, Weidmann JL, Prins LJ, Cardullo F, Snellink-Ruël BH, Fokkens RH, Nibbering NM, Shinkai S, Reinhoudt DN. Ag+ labeling: a convenient new tool for the characterization of hydrogen-bonded supramolecular assemblies by MALDI-TOF mass spectrometry. *Chemistry* 2000, 6:4104-4115.
151. Turkewitz AP, Kelly RB. Immunocytochemical analysis of secretion mutants of Tetrahymena using a mucocyst-specific monoclonal antibody. *Dev Genet* 1992, 13:151-159.
152. Valet G. Cytomics: an entry to biomedical cell systems biology. *Cytometry A* 2005, 63:67-68.
153. van Leeuwen MB, Deddens AJ, Gerrits PO, Hillen B. A modified Mallory-Cason staining procedure for large cryosections. *Stain Tech* 1990, 65:37-42.
154. van Tongeren M, Cherrie JW. An integrated approach to the exposome. *Environ Health Perspect* 2012, 120:A103-A104.
155. Vieth M, Sutherland JJ, Robertson DH, Campbell RM. Kinomics: characterizing the therapeutically validated kinase space. *Drug Discov Today* 2005, 10:839-846.
156. Vogler N, Meyer T, Akimov D, Latka I, Krafft C, Bendsoe N, Svanberg K, Dietzek B, Popp J. Multimodal imaging to study the morphochemistry of basal cell carcinoma. *J Biophot* 2010, 3:728-736.
157. von der Lieth CW, Bohne-Lang A, Lohmann KK, Frank M. Bioinformatics for glycomics: status, methods, requirements and perspectives. *Brief Bioinform* 2004, 5:164-178.
158. Von Foissner W. Morphologie und Infraciliatur einiger neuer und wenig bekannter terrestrischer und limnischer Ciliaten (*Protozoa, Ciliophora*). Sitzungsberichte der Osterreichischen Akademie der Wissenshaften. Mathematisch-Naturwissenschaftliche Klasse 1989, Abt. I, 196:173-247.

159. Vrijheid M. The exposome: a new paradigm to study the impact of environment on health. *Thorax* 2014, 69:876-878.
160. Wang L, Xie XQ. Computational target fishing: what should chemogenomics researchers expect for the future of in silico drug design and discovery? *Fut Med Chem* 2014, 6:247-249.
161. Wasik A, Mikołajczyk E. Infraciliature of Cymatocylis affinis/convallaria (Tintinnina). *Acta Protozoologica* 1994, 33:79-85.
162. Welsh MG. Current methodologies for the study of pineal morphophysiology. *J Pineal Res* 1994, 16:113-120.
163. Wenk MR. The emerging field of lipidomics. Nat Rev Drug Discov 2005, 4: 594-610.
164. Werner T. Proteomics and regulomics: the yin and yang of functional genomics. *Mass Spectr Rev* 2004, 23:25-33.
165. Westbrook JA, Yan JX, Wait R, Dunn MJ. A combined radiolabelling and silver staining technique for improved visualisation, localisation, and identification of proteins separated by two-dimensional gel electrophoresis. *Proteomics* 2001, 1:370-376.
166. Wilbert N. Ciliaten aus dem Interstitial des Ontario Sees. *Acta Protozool* 1986, 25:379-396.
167. Willard HF. Chromonomics. Centromere structure and function in natural and artificial human chromosomes. Cytogenet Cell Genet 1999, 85:6.
168. Winkler C, Denker K, Wortelkamp S, Sickmann A. Silver- and Coomassie-staining protocols: detection limits and compatibility with ESI MS. *Electrophoresis* 2007, 28:2095-2099.
169. Wuster A, Madan BM. Chemogenomics and biotechnology. *Trends Biotechnol* 2008, 26:252-258.
170. Yagiu R, Shigenaka Y. Electron microscopical observations of Condylostoma spatiosum in ultra-thin section. IV. The fibrils between the basal granule and the longitudinal fibrillar bundle. *Zool Mag* 1959, 68:414-418.
171. Yasuda K. On-chip cellomics assay enabling algebraic and geometric understanding of epigenetic information in cellular networks of living systems. 1. Temporal aspects of epigenetic information in bacteria. *Sensors* 2012, 12:7169-7206.
172. Zaman V. The application of fluorescent antibody test to Balantidium coli. *Trans Roy Soc Trop Med Hyg* 1965, 59:80-82.
173. Zaman V. The application of fluorescent antibody test to Balantidium coli using fluorescent isothiocyanate and rhodamine isothiocyanate. *Med J Malaysia* 1966, 20:325.
174. Zhang W, Li F, Nie L. Integrating multiple 'omics' analysis for microbial biology: application and methodologies. *Microbiology* 2010, 156:287-301.
175. Zhang WJ, Lin YS, Cao WQ, Yang J. Genetic diversity and variance of Stentor coeruleus (*Ciliophora: Heterotrichea*) inferred from inter-simple sequence repeat (ISSR) fingerprinting. *J Eukaryot Microbiol* 2012, 59:157-162.
176. Zhang WJ, Yang J, Yu YH, Shu SW, Shen YF. Population genetic structure of Carchesium polypinum (*Ciliophora: Peritrichia*) in four Chinese lakes inferred from ISSR fingerprinting: high diversity but low differentiation. *J Eukaryot Microbiol* 2006, 53:358-363.
177. Zhu J, Ofek G, Yang Y, Zhang B, Louder MK, Lu G, McKee K, Pancera M, Skinner J, Zhang Z, Parks R, Eudailey J, Lloyd KE, Blinn J, Alam SM, Haynes BF, Simek M, Burton DR, Koff WC; NISC Comparative Sequencing Program, Mullikin JC, Mascola JR, Shapiro L, Kwong PD. Mining the antibodyome for HIV-1–neutralizing antibodies with next-generation sequencing and phylogenetic pairing of heavy/light chains. *Proc Natl Acad Sci USA* 2013, 110: 6470-6475.
178. Zwickl H, Traxler E, Staettner S, Parzefall W, Grasl-Kraupp B, Karner J, Schulte-Hermann R, Gerner C. A novel technique to specifically analyze the secretome of cells and tissues. *Electrophoresis* 2005, 26:2779-2785.